AMERICA UNDERGROUND

AMERICA UNDERGROUND

Christie McFall

Illustrated with photographs and with drawings and maps by the author

COBBLEHILL BOOKS
Dutton • New York

To my wife, Olga, who made this book possible

Copyright © 1992 by Christie McFall
All rights reserved
No part of this book may be reproduced in any form
without permission in writing from the publisher.

Library of Congress Cataloging-in-Publication Data
McFall, Christie.
  America underground / Christie McFall ; illustrated with
photographs and with drawings and maps by the author.
    p. cm.
  Includes index.
  Summary: Describes all the things found underground, including living creatures, caves, minerals, tunnels, subways, offices, and landfills.
  ISBN 0-525-65079-2
  1. Underground construction—Juvenile literature. 2. Underground areas—United States—Juvenile literature. [1. Underground construction. 2. Underground areas.] I. Title.
TA712.M39 1992
624.1′9′0973—dc20   91-8951   CIP   AC

Published in the United States by Cobblehill Books,
an affiliate of Dutton Children's Books,
a division of Penguin Books USA Inc.
375 Hudson Street, New York, New York 10014

Designed by Susan Phillips
Printed in the United States of America
First Edition 10 9 8 7 6 5 4 3 2 1

CONTENTS

1. LOOK DOWN 7
2. LET'S GO DEEPER 14
3. DARK CAVERNS 26
4. RECOVERING RICHES 35
5. WORKING UNDERGROUND 45
6. PIPELINES AND PASSAGEWAYS 53
7. UNDER CITY STREETS 62
8. IT HAS TO GO SOMEWHERE 72
   INDEX 78

1
LOOK DOWN

An amazing world lies beneath you. As you walk down a dirt path, do you ever think about what may be under your feet? The soil is laced with roots and is teeming with tiny creatures, most of them invisible to the naked eye.

If you were to take a shovel and peel back a layer of sod, you would see a tangled mat of roots. An incredible number of roots are packed into our soil, most of them in the upper few feet. In a well-known experiment in the 1930s, a researcher planted a single winter rye seed in a box of soil. After four months' growth, he removed the rye plant and carefully washed the soil away. Then he painstakingly measured each strand of the plant's dense mass of roots. He found that if all the roots were laid end to end, they measured an astonishing 387 miles!

There are two principal kinds of roots—*tap roots* and *fibrous roots*. Grasses, like winter rye, have fibrous roots, which have a mass of fine roots of approximately the same size branching out from the base of the stem. Each of those roots divides into many thousands of smaller roots, which in turn branch into even finer root systems until the plant has millions of tiny roots. However, this amazing root system is not easily observed, because when a plant is pulled out of the earth, most of the finer fibrous roots remain in the ground attached to soil particles.

Tap roots are large primary roots that reach deep into the soil and anchor the plant securely. Large trees such as pine and shagbark hickory have tap roots. Secondary roots growing out from the tap root are smaller in diameter but are widespreading, often extending well beyond the width of the tree's branches. Many root systems are enormous, equaling or exceeding in size the part of the plant that is above ground. Growing tree roots can exert enough pressure to split boulders or to buckle sidewalks, as you have probably seen.

SOIL ORGANISMS

The type of soil in a particular location helps determine the kinds of plant life growing there, as well as the kinds of animal life living in the soil. Soil is made up of mineral particles—clay, silt, and sand combined with decayed vegetable and animal matter (known as *humus*). There are thousands of different soils in the United States, ranging from wet, black, mucky soils to sandy desert soils.

Is the earth solid? Not really. Spaces between the mineral particles that make up soil contain air, water, or both, and make up 30 to 60 percent of soil volume, depending on the coarseness of the soil. These spaces harbor a multitude of living things. A cubic inch of soil can contain billions of microscopic creatures—bacteria, fungi, viruses, and protozoans. Bacteria are so small that 250,000 could fit on a pinhead. Some are harmful to humans, others are helpful. One kind causes the disease, tetanus, but other soil organisms have been used in the development of penicillin and streptomycin, two life-saving drugs. Perhaps more importantly, the bacteria and fungi in the soil break down the dead plant and animal matter into molecules that plants can use for food. Roundworms (*nematodes*) which feed on roots are found in all soils. Though smaller than a comma on this page, they are thousands of times larger than bacteria—just barely visible to the naked eye. Under a microscope, nematodes resemble tiny eels.

ANIMALS IN THE SOIL

Within the average lawn there may live hundreds of species of animals large enough to be seen. The soil is honeycombed with the passageways and burrows of insects, worms, white grubs, and centipedes. Earthworms are probably the best known of the soil dwellers. As earthworms tunnel through the soil, they draw soil and vegetable matter into their mouths, pass it through their bodies, and deposit it as rich, black "castings" in their wake. Earthworms in a single acre may pass some 40 tons of soil through their bodies each year.

About 95 percent of all insect species (nearly a million) inhabit the soil at some time in their life cycles. Many insects deposit their eggs in the soil. Grasshoppers do, and some species of bees tunnel into the soil to build nests and lay eggs. But no insect relies more on the soil for incubation than the cicada, or 17-year locust. The $\frac{1}{16}$ inch-long nymph breaks out of its egg case on the bark of a tree, falls to the earth, digs in four to six inches below ground level, and feeds from the sap of tree roots through a sucking beak. The nymph matures over a 17-year period, then burrows upward through the soil, stopping about an inch below the surface. When it emerges from underground it is transformed into an adult winged cicada within hours. The females deposit eggs six weeks later and the cycle begins again. In southern states the cicada's life cycle is 13 years.

LIFE IN THE SOIL

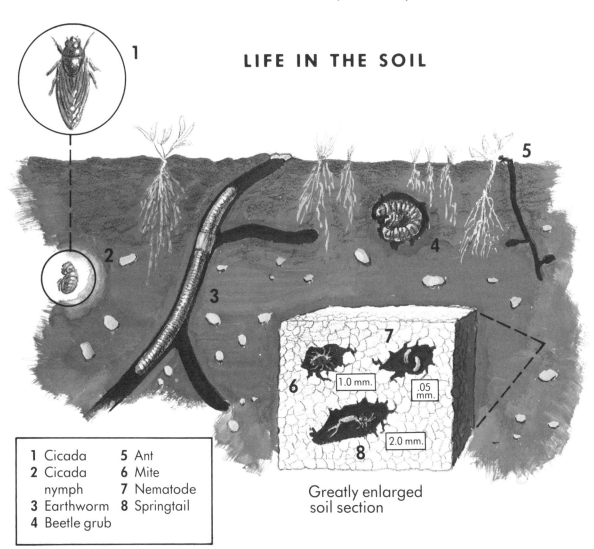

1 Cicada
2 Cicada nymph
3 Earthworm
4 Beetle grub
5 Ant
6 Mite
7 Nematode
8 Springtail

Greatly enlarged soil section

FOOD FROM UNDERGROUND

Many of the foods we eat are the roots or underground stems of plants. Carrots and beets are roots; potatoes are swollen underground stems (tubers) that develop in the soil. Anyone who has grown potatoes in a backyard garden knows the fun of reaching into the loose soil to discover one potato after another.

American Indians ate many roots and underground stems. One of their favorites was the groundnut, which resembles a small potato. The Wampanoag Indians shared groundnuts with the Pilgrims, helping the colonists to survive their first difficult years. Captain John Smith, of the Jamestown colony, spoke of "Groundnuts as big as egges and as good as potatoes, and 40 on a string, not two inches underground."

BURROWING ANIMALS

Animals are also attracted to the tasty root crops. Rodents, the most numerous order of mammals, can exist even in desert areas because of the many plants which have fleshy roots, bulbs, or tubers.

Small openings in the earth, often hidden in brush or weeds, are usually the only sign of networks of tunnels made by burrowing animals. They dig passageways to hunt their prey, to rear their young, to take shelter from predators, and to hibernate. The most remarkable excavations were made years ago by prairie dogs in the Great Plains. A single underground town once spread out over an area 100 miles wide by 250 miles long. It contained an estimated 400 million inhabitants.

One of our most common burrowing animals is the woodchuck or groundhog. Its heavyset body is about two feet long

GROUNDHOG BURROW

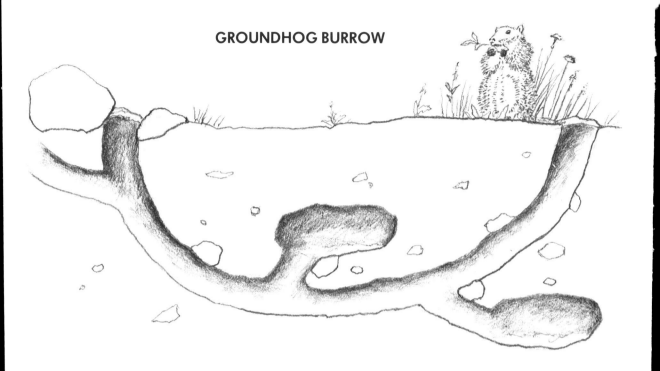

and covered with thick, coarse hair. It nests in burrows that may have several compartments, and hibernates in winter, its heartbeat and breathing slowing down. The groundhog is said to leave its burrow on February 2, and to return underground if it sees its shadow. If it does, we supposedly have six more weeks of winter.

A great nuisance to gardeners is the common mole, which often uproots bulbs and other garden plants in its underground travels. Raised passageways can be easily seen winding across lawns where the mole has tunneled one or two inches below the surface. More heavily used burrows are dug a foot or more beneath ground level, and dirt is forced to the surface in a series of "molehills." A single mole can tunnel as far as 200 feet in a day.

Unfortunately, it is seldom possible to observe the fascinating subterranean world firsthand. However, the Arizona Sonora Desert Museum in Tucson has constructed a 175-foot tunnel in the desert which allows visitors to observe animals and plant roots of the Sonoran Desert from underground.

2
LET'S GO DEEPER

      Beneath the soil are rocks. In fact, most of the earth's crust is made up of rocks. There are different kinds, depending on the way they are formed. *Sedimentary rocks* are formed when older rocks are worn away into tiny pieces that are carried off by running water, wind, glaciers, or gravity. This material (*sediment*) is then deposited in rivers, lakes, and deltas. All sedimentary rocks are built up in layers and in time the weight of additional deposits compresses them into solid rock.

   Igneous rocks are those formed by the cooling and solidifying of *magma*, hot liquid rock deep within the earth. The North American continent—and other continents—are largely granite, an igneous rock. *Metamorphic* rocks are those formed when sedimentary or igneous rocks are changed by intense heat, great pressure, or chemical reactions.

Solid rock underlying the soil is called *bedrock*. Although bedrock appears to be solid, it is capable of holding great quantities of water. There are tiny pore spaces in rocks which fill with water from rain or melted snow. This water within rocks is known as *groundwater*. Pores and cracks in rocks at great depths under the earth's surface virtually close because of the weight of overlying rocks, leaving no pore space for water. But there is 20 to 30 times as much water underground as there is in all our lakes, rivers, and streams combined.

Why is groundwater important? Because it is the source of water for irrigation, for drinking, and for industrial use. Groundwater is preferred to surface water because its temperature remains constant, and it is generally free of pollution. The top level of groundwater is called the *water table*. It may be near the surface in humid regions, during wet seasons, or in valley bottoms. In some arid parts of the United States, water is being pumped that fell as rain during the Ice Age 10,000 years ago.

AQUIFERS

A layer of rock, sand, or gravel that has a usable supply of groundwater is called an *aquifer*. Gravel, limestone, and sandstone are the best carriers of water. Wells in such regions yield moderate to large quantities of water.

One of our most important aquifers is the broad coastal plain from Cape Cod to the Mexican border. Most of the area consists of sand (not yet compressed into sandstone) confined by layers of clay. Water is readily drawn from sand and gravel. Florida differs from the rest of the coastal plain. It has limestone bedrock that forms a vast subterranean storage area for ground

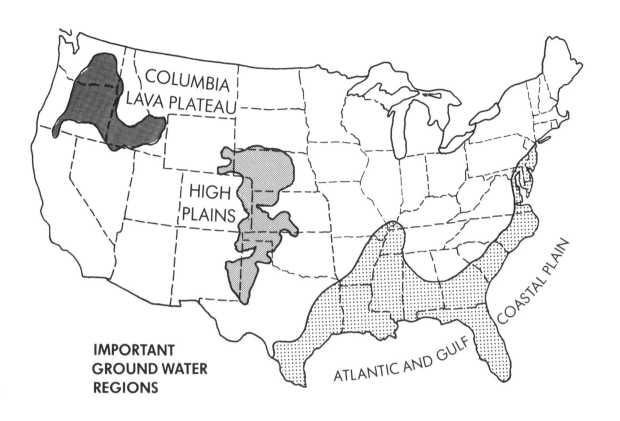

IMPORTANT GROUND WATER REGIONS

water. In southern New Jersey, beneath the Pine Barrens, the aquifer holds as much water as a 1,000-square-mile lake 75 feet deep. The groundwater of the Pine Barrens approaches the purity of uncontaminated rainwater or melted glacial ice, according to the U.S. Geological Survey. The high acidity of the water prevents bacterial growth—a quality greatly appreciated by many seamen of the nineteenth century who carried this water with them on long ocean voyages because it stayed pure for months.

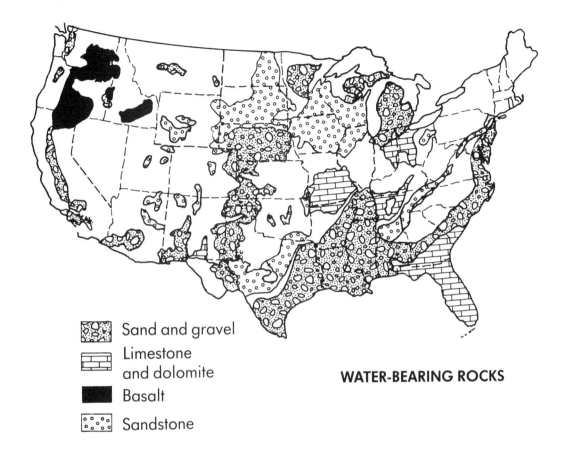

- Sand and gravel
- Limestone and dolomite
- Basalt
- Sandstone

WATER-BEARING ROCKS

Another large and important aquifer is the Ogallala, which lies under parts of eight states, from Nebraska down into New Mexico and Texas. Farmers of Texas have transformed a barren, dust-bowl landscape into one of the richest agricultural regions in the world—but at a price. The water table is dropping alarmingly. In parts of Texas and Kansas it is falling as much as two to five feet a year. In many areas the aquifer is nearly drained. The desert floor between Tucson and Phoenix in Arizona is laced by cracks as much as 25 feet wide and 400 feet deep.

Lava cliff with stream flowing over it and groundwater emerging from it

SPRINGS AND GEYSERS

Water coming to the surface from the ground or from rocks results in *springs*. What would seem an unlikely spot for groundwater to emerge is the great Columbia River plateau in eastern Washington and Oregon and parts of Idaho. Here, thick lava beds alternate with sand and other deposits. Great springs along the Snake River canyon form one of the most spectacular displays of ground water discharge in the world.

More than 1,100 springs can be found in Missouri. The ten largest have a combined flow of over a billion gallons of water on an average day. Another area where notable springs are found is in Florida. As the Suwannee River flows out of Okefenokee Swamp across northern Florida and into the Gulf of Mexico, fifty springs bubble up along its valley. The water that gushes forth is crystal clear. Wakulla Spring in Florida pours out the greatest volume from a single fissure of any spring on earth.

Some springs are warm, or even hot, fed by water heated by hot rocks far below the surface. *Geysers* are thermal (hot) springs that erupt violently, intermittently, and to varying heights, throwing hot water and steam into the air. Yellowstone National Park in Wyoming contains one of the world's greatest collections of hot springs and geysers. There are two hundred geysers—including Old Faithful—and thousands of hot springs.

EARTHQUAKES AND VOLCANOES

Two of the most awesome natural occurrences, volcanic eruptions and earthquakes, have their origins underground. Volcanoes sometimes erupt with explosive force. Earthquakes shake

Steam vents, Yellowstone National Park

the very earth beneath our feet and may precede volcanic eruptions as the bedrock shifts suddenly along a split or fracture in the earth's crust.

The power of an earthquake to shift monstrous blocks of land in a few minutes was demonstrated on March 27, 1964, when Alaska was struck by one of the strongest earthquakes of the twentieth century—8.5 on the Richter scale. A region the size of Maine was uplifted an average of seven feet! In one case, a mother and her two children ran from their house as the earth of their cliffside lot began buckling and great cracks appeared. They clung to a huge, tilting block of earth as it slid down to sea level. Miraculously, they survived.

Most earthquakes in the United States occur along the Pacific coastline. One of the most active areas is the San Andreas fault in California, which marks the boundary between two of the earth's great shifting crustal plates. While the American Plate remains fairly steady, the Pacific Plate is moving northward. During the famous San Francisco earthquake of 1906, the earth in places shifted as much as 21 feet.

More recently, on October 17, 1989, an earthquake that registered 7.1 on the Richter scale jolted the San Francisco Bay area, taking lives, causing widespread damage, and shaking up spectators in Giant Stadium who were about to see the third game of baseball's World Series. The buildings that collapsed in San Francisco were built on soft, unstable soil, and were unreinforced buildings. Strict adherence to San Francisco building codes, the strictest in the nation, saved thousands of lives and hundreds of buildings.

School that was wrecked during Alaska earthquake

VOLCANIC ERUPTIONS

Volcanoes have their beginnings in molten rock, which lies only a few miles under the earth's surface. This hot liquid rock may rise to the surface through a vent in a volcano and erupt violently into the air, or it may pour out slowly through a long fracture and spread out over the land. When underground, this molten rock is known as *magma*. When it reaches the surface, it is called *lava*.

Mount St. Helens in the state of Washington has been more active and more exoplosive than any other volcano in the lower forty-eight states in recent years. In the spring of 1980, it was a beautiful snow-capped cone, but pent-up forces were at work below the surface. Scientists noticed the first signs of a type of earthquake tremor that signals the underground movement of molten rock. On the morning of May 18, the mountain exploded. A choking ashfall blanketed the area. The volcano hurled out 275 million tons of superheated ash and rock, igniting widespread fires. Mud and debris swept away everything in its path.

ALASKA AND THE HAWAIIAN ISLANDS

A chain of volcanic islands 1,500 miles long form Alaska's Aleutian Islands. Of the 76 major volcanoes in this chain, 36 have erupted since 1760, making it one of the world's most active volcanic regions.

The Hawaiian Islands are the tops of giant volcanic mountains that rise out of the Pacific Ocean. They were formed over

Mount St. Helens erupting

Observatory is very near Kilauea volcano's cauldron rim.

several million years through countless outpourings of extremely fluid lava, which formed very broad dome-shaped *shield* volcanoes. The island of Hawaii, made up of five separate volcanoes, merged to form the highest volcanic mass on earth, rising nearly six miles above the ocean floor.

The Hawaiian volcanoes have drawn worldwide attention because of frequent spectacular eruptions which can be viewed and studied with relative ease and safety. Kilauea, located on the slopes of Mauna Loa, is the most active volcano in the world. The Hawaiian Volcano Observatory is near Kilauea cauldron's rim. It is both accessible and safe—safe, because Hawaii's volcanoes do not erupt with explosive force. They do pour out lava periodically, adding to the islands' height and extending the shoreline. Lava called *aa* (ah ah) has a rough rocky crust; *pahoehoe* (pah ho-ee ho-ee) is ropy with a shiny surface.

In September, 1977, Kilauea erupted, sending a fiery river of lava 40 feet deep and 1,000 feet wide heading for the tiny village of Kalapana. It stopped short just 400 yards away. Kilauea began its last eruptive phase six years later. Since then, most of Kalapana had been overrun by lava. The last remaining landmarks—the Kalapana store and drive-in—were destroyed in 1990.

Pahoehoe lava drapes over a sea cliff.

3
DARK CAVERNS

Beneath large regions of America lies a subterranean world of tunnels, shafts, passages, and underground rivers. It is a black world, blacker than the blackest paint or the darkest night, a world of total darkness. These chambers within the earth are caves. There are some 17,000 known caves in the United States.

A cave is a natural underground chamber open to the surface. It extends beyond the zone of light and is large enough for a person to enter. Caves are formed in many different materials—gypsum, salt, marble, lava—but caves are most commonly found in limestone. Most of the world's caves have been formed over hundreds of thousands of years by the slow dissolving of limestone. As ground water seeps downward through decaying plant and animal matter, a chemical reaction takes place that dissolves the limestone.

Scientists believe that the first stage in cave development takes place just below the water table where the porous rock is saturated with water. The underground water slowly dissolves some of the rock, leaving passages and chambers. The second stage takes place if the water table drops below the level of the cave. This leaves passageways in the unsaturated zone where air from above can enter. If water from the surface continues to seep through cracks in the rock above these passageways, some of the dissolved minerals in the water crystallize and are deposited within the cave in a variety of fantastic forms which are called *speleothems*—cave formations or decorations.

SPELEOTHEMS

Water dripping from cave ceilings forms mineral deposits called *stalactites*. They often begin as narrow tubes, only one drop in width. As the "soda straw" elongates, it eventually clogs and the mineral-laden water begins to drip down the sides, forming an icicle shape. Each stalactite grows at a different rate, depending on the wetness of the cave, the temperature, and the thickness of the rock layer above the cave. Some stalactites grow a few centimeters a year, while others may take hundreds of years to grow that much.

When drops splash on the cave floor, thick *stalagmites* slowly build up toward the ceiling. Stalactites and stalagmites may grow together to form a column. Stone *draperies* are formed where water has run down a slanted ceiling. Water flowing down a wall or over the cave floor builds up layers of *flowstone*.

MAMMOTH CAVE NATIONAL PARK

Caves are usually horizontal. A good example is Mammoth Cave in Kentucky, the world's longest known cave. Its various passageways on several levels exceed 330 miles in length.

Central Kentucky is honeycombed with *sinkholes*—usually funnel-shaped depressions. If you were to drive through the Mammoth Cave region, you might wonder how one could travel for miles through this green and obviously well-watered land without ever crossing a stream. The water is there, but it is mostly underground. At least twenty-one streams vanish into sinkholes or "swallow-holes." Because of the 60,000 sinkholes in Mammoth Cave National Park, water drains from the surface very quickly, disappearing into the ground as if through a sieve. The Echo River, 360 feet below the surface, travels underground through the lowest of the five levels in Mammoth Cave, and drains into the Green River.

Mammoth Cave contains many large chambers connected by narrow passages. Beautiful stalactites, stalagmites, and columns have formed. Bats and insects are found, and there are eyeless fish in the lakes or rivers.

CARLSBAD CAVERNS NATIONAL PARK

The limestone caves of Carlsbad Caverns are found in New Mexico in the foothills of the Guadeloupe Mountains. Here, too, there are even more fantastic stalactite and stalagmite formations. There are three distinct levels, with the most extensive passages being in the middle level, 750 feet below the surface. One of the largest underground chambers in the world is the Big Room—225 feet high, 1,800 feet long, and up to 1,100 feet

Natural entrance to Carlsbad Caverns

wide. The walkway around the perimeter of the Big Room covers one and a quarter miles.

Carlsbad Caverns was discovered in 1901 by Jim White, a cowboy, who was drawn to the site when he saw a swirling column of bats in the sky. Every evening at dusk during the warm months, visitors assembled at the entrance to watch as some 300,000 Mexican free-tailed bats fly out of the cave to the desert to feed on insect life during the night. At dawn the returning bats fly high above the cave entrance, fold their wings, and plummet back into the cave. The bats attach themselves

to ceilings and walls in a part of the cave less than 200 feet down, but do not invade the corridors used by visitors, which are at lower levels. As many as 300 bats may crowd into one square foot of ceiling. The bats stay in Carlsbad from early spring through October, then winter in Mexico.

There are more than seventy other caves in the park. One of the newest is Lechuguilla Cave, only a few miles from Carlsbad Caverns. Although this cave has been known since 1914, it was not seriously explored until 1987. Today, it is America's deepest cave and the fourth longest. Experienced cavers have been overwhelmed by its rare and amazing formations.

Jewel Cave National Monument in South Dakota has more than 73 miles of passageways and rooms lined with calcite crystals which sparkle like gems. Two private caves, the Caverns of Sonora, Texas, and Luray Caverns in Virginia, are among America's most beautiful caves. Both are designated Natural Landmarks by the National Park Service.

LAVA CAVES

Although limestone caves take hundreds of thousands of years to develop, lava caves are formed in just days or weeks. Such caves occur where lava has recently flowed from volcanoes. The surface solidifies while the lava underneath remains molten and continues to flow. Underground streams of molten rock may continue to course through slowly hardening channels and drain away in the final stages of the eruption, leaving tubular caves. The largest known lava cave, Kazikura Cave in Hawaii, is seven miles long.

Giant's Hall is one of the highlights of Luray Caverns.

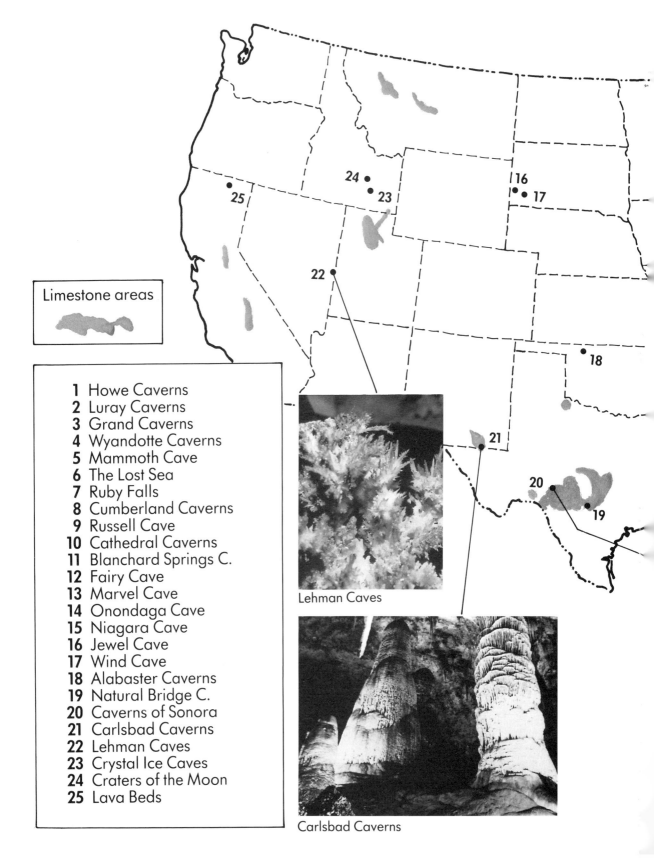

Limestone areas

1 Howe Caverns
2 Luray Caverns
3 Grand Caverns
4 Wyandotte Caverns
5 Mammoth Cave
6 The Lost Sea
7 Ruby Falls
8 Cumberland Caverns
9 Russell Cave
10 Cathedral Caverns
11 Blanchard Springs C.
12 Fairy Cave
13 Marvel Cave
14 Onondaga Cave
15 Niagara Cave
16 Jewel Cave
17 Wind Cave
18 Alabaster Caverns
19 Natural Bridge C.
20 Caverns of Sonora
21 Carlsbad Caverns
22 Lehman Caves
23 Crystal Ice Caves
24 Craters of the Moon
25 Lava Beds

Lehman Caves

Carlsbad Caverns

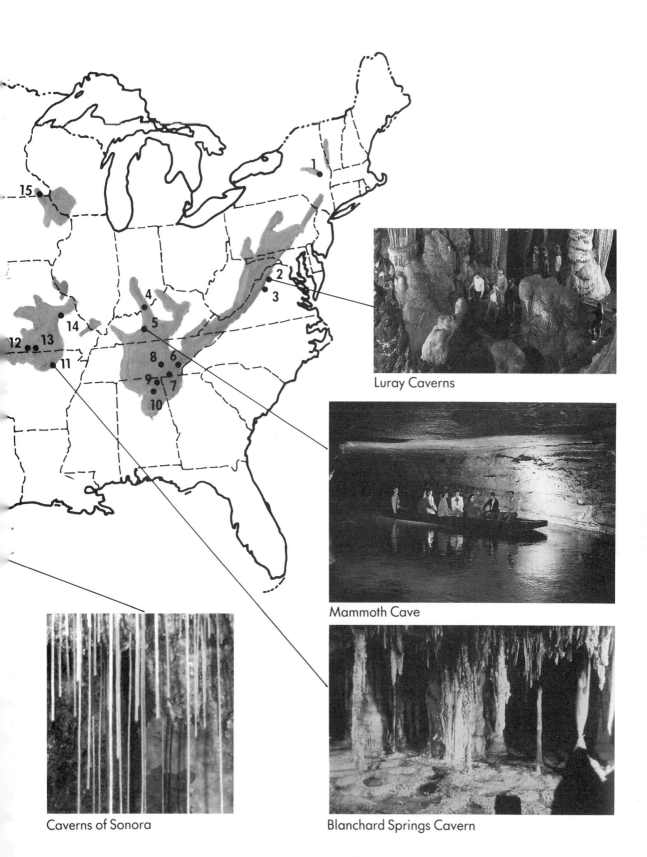

Luray Caverns

Mammoth Cave

Caverns of Sonora

Blanchard Springs Cavern

A network of tunnels underlies 72 square miles of Lava Beds National Monument in California, which contains the world's greatest concentration of lava tubes or caves—at least 300. Some have four or five levels.

SINKHOLE CAVES

The submerged caves in Florida are water-filled sinkholes. Most of Florida's caverns lie below the water table, which is close to the surface. To explore these water-filled caves, divers must use scuba equipment. These cave explorers can cover long passages with relative ease, but then they often lose their way in the murky waters, even with powerful lights. Jagged cave formations are always a threat to sever a diver's air hose. In a twenty-year period, 234 people lost their lives in these dangerous caves. But in spite of the dangers and difficulties in all forms of caving, adventurers across America continue to search for new caves to explore and for new passages in known caves.

OPEN TO THE PUBLIC

There are about 125 caves in America which are open to the public. Most have hidden lighting and easy walkways. Many have elevators, and Mammoth Cave even has a 1½-mile tour just for people in wheelchairs. Temperatures in a cave are usually cool—the same as the average annual temperature at the surface. Humidity is usually nearly 100 percent.

Each cave has its own personality. Some have draperies, cave pearls, gypsum needles, or other fanciful formations. There are no "two of a kind."

4
RECOVERING RICHES

America is blessed with a wealth of underground riches, especially oil, natural gas, and coal, our three main energy sources. Numerous minerals also lie below the surface, and half the water we use is drawn from underground sources. Some of these riches are dug out near the surface; others may take miles of drilling to recover.

OIL AND NATURAL GAS

Oil is vital to America. It heats our homes, runs our cars, and is used in the manufacture of many products. Today, the United States ranks second in the world among oil producers, after the Soviet Union.

Natural gas is a flammable gas issuing from natural openings in the earth's crust or obtained in the production of petroleum. It is second only to crude oil in importance as a fuel. Natural gas flows from 169,000 wells in the United States.

DRILLING WELLS

When drilling for oil, a sharp drilling bit is rotated at the end of a "string" of pipe. In a deep well, a massive hook on the drilling rig may have to support the weight of more than four miles of steel drill pipe. In many drilled wells, sections of steel pipe called *casing* are inserted into the hole, extending to the full depth of the well. The casing prevents rocks and dirt from breaking away from the sides of the hole and clogging the opening made by the drilling tool.

Generally, the temperature at the bottom of the hole rises between one and two degrees Fahrenheit for every 100 feet in depth. The deeper the rock, the more dense it becomes, so the rock is harder to drill and bits need to be replaced more often. In a 20,000-foot well, it takes some ten to twenty hours to pull the entire string of drill pipe to the surface, replace the bit, and reset the drill pipe.

Much of our oil production takes place on the Continental Shelf in the Gulf of Mexico. Here, some 100 wells are located, drilled from platforms and drill ships. As many as 62 wells can be drilled into the sea bottom from a single platform.

Underground water is withdrawn from wells that reach down below the water table. If the water table is near the surface, shallow wells can be dug. Deep wells must be drilled with special equipment similar to that used for oil. When the hole has been drilled to some distance below the top of the

Workers on an oil derrick

water table, a water pipe is lowered inside the casing. The lower end of this pipe has many perforations that allow water to enter but exclude sand or dirt. Water is forced up the pipe by means of a motor-driven pump.

COAL

In addition to oil and natural gas, America's other great source of energy is coal. We have more than 5,000 coal mines, but the fifty largest companies produce two-thirds of our nation's coal.

Most coal is *bituminous*—a soft coal. *Anthracite* is a jet black coal with a high luster. In the United States it is found only in deep underground deposits in northeast Pennsylvania. This hardest of all coals is formed under extreme pressure.

By the 1980s, more coal was being extracted from surface strip mines than by tunneling underground. Gigantic surface-

Continuous mining machine digs coal from underground horizontal seams in a single operation.

mining shovels, as tall as a 20-story building, strip off soil and rock to uncover buried coal seams. The shovels scoop up more than 200 tons in a single bite.

Deep mines are found mostly in the East. The Appalachian plateau, from northern Pennsylvania to Alabama, is the chief source of coal. In this region of steep-walled valleys, most of the mines are underground. Some of the larger mines have several hundred miles of tunnels.

Some underground coal is still removed by blasting, but two-thirds of the coal from mines is removed by machines which gouge it out and load it onto shuttle cars or a conveyor belt.

Loading machine sweeps lumps of coal onto conveyor which drops them into a shuttle car.

MINE SAFETY

Fires, floods, and cave-ins have taken the lives of thousands of miners. It took a series of disasters to focus federal attention on health and safety in mining. In 1907, explosions tore through two coal mines in West Virginia, killing over 400 men. Thirteen days later, more than 200 miners were killed in Pennsylvania. In the years from 1904 through 1930, 64,442 miners died in accidents.

"Black lung" is contracted by miners from years of breathing coal dust, and is often fatal. And in the early days there were no child labor laws. Eight-year-old boys earned twenty-five cents a day picking slate out of the coal after it was crushed.

There are several dangerous gases found in coal mines. *Methane*, or marsh gas, forms, and since it is lighter than air, it accumulates near the roofs of tunnels. It is tasteless, colorless, and odorless, and when mixed with air it is violently explosive—the cause of many mining disasters. Coal dust igniting in the air can produce explosions as violent as those produced by gases.

Improvements in mine safety have sharply reduced the death rate from mine accidents. Today, air vents with fans at the surface prevent dangerous gases from accumulating, and methane detectors are required by federal law. Powdered limestone is sprayed on walls to dilute the coal dust.

ABANDONED MINES

Danger from coal mines does not stop when they finally close down. There are abandoned mine shafts all across America. The settling of rock and earth overlying abandoned mines can damage streets and buildings above, causing cave-ins or cracks in structures. This happens when rock above the mine had been inadequately supported.

FIRES UNDERGROUND

Worse than the settling of earth and rock over abandoned coal mines are the 250 underground coal mine fires burning across the United States. Theodore Roosevelt wrote about burning coal seams in the 1880s at what is now Theodore Roosevelt National Park in North Dakota. Exposed veins are usually started burning by range fires or lightning. Several are burning today.

Perhaps the worst of these underground fires is now burning below Centralia, Pennsylvania. Here, an inferno is consuming a thick bed of anthracite coal at temperatures that approach 2000 degrees Fahrenheit—a temperature that will practically melt rock. The fire has burned for more than twenty-five years. In places, the surface is so hot (1000 degrees F.) that a piece of paper dropped on the ground will turn instantly into white ash. In recent years most of the town has been relocated.

MORE WEALTH

Other mineral substances are taken from the earth, as different as gold and gravel, stone and salt, lead and limestone. One important mineral that provides a source of energy for nuclear power plants is uranium. More than a quarter of the world's supply of uranium underlies our western states, most of it in New Mexico and Wyoming.

Many metals are found underground. Some 4,000 years ago,

Indians of northern Michigan were mining rich copper deposits. To extract the copper, they built bonfires in pits they dug and poured cold water over the heated rocks, splitting them. Then they beat out the copper with stone hammers.

A mining district extending along the Montana-Idaho line has been producing lead, zinc, silver, and gold continuously since the 1880s. Four of the largest lead mines, two of the ten largest zinc mines, and four of the largest silver mines are in this district.

GOLD

The discovery of gold in the mid-1800s began the "gold rush" to the American West of at least a quarter million men, seeking their fortunes. It is estimated that there may have been as many as 100,000 mining districts by 1900. Most were small and short-lived, but a few mines struck it rich, one with an output of a billion dollars worth of gold. When the ore ran out, most camps and settlements were abandoned to become "ghost towns."

Today, there is gold mining in nearly all of the western states and in Alaska. Recently, rich gold deposits have been discovered in the Tuscarora Mountains north of Elko, Nevada. This is micro-gold, only visible through an electron microscope. The rock contains an average .04 of an ounce of gold per ton.

DIAMONDS

Diamonds have been found and mined commercially at only one place in America. It is now part of Crater of Diamonds state park in Arkansas. For a fee, visitors to the park can hunt for diamonds in the weathered rock. An average of a diamond a day is found, but the vast majority are quite small in size.

GEOTHERMAL ENERGY

One other great but largely untapped source of energy lies mostly west of the Rocky Mountains—geothermal energy (steam and hot water.) All along the world's earthquake and volcano belts, pockets of magma or molten rock have worked their way close to the surface. Water in the permeable rock layers above these pockets is heated, often far above the normal boiling point. Sometimes the superheated water forces its way out as hot springs or geysers. More often, trapped by a cap rock above it, the water becomes a subterranean reservoir, which can

The Geysers geothermal plant near San Francisco

be trapped by wells and the water harnessed to drive turbines of electrical generators.

The Geysers, 90 miles northeast of San Francisco, is the largest geothermal facility in the world. The generating plant's steam-driven turbines produce enough power for a city half the size of San Francisco. In Boise, Idaho, hot springs have heated homes since the 1890s. Several state buildings are also using geothermal energy.

5
WORKING UNDERGROUND

Limestone is used extensively for buildings. It is quarried from open excavations when it is near the surface, but it is also mined by tunneling underground. Today, after the limestone has been removed, some of these mines are being given new life as warehouses, factories, and offices.

Missouri leads the nation in the use of mined-out space, with the greatest concentration being under Kansas City, which has at least twenty-eight sites. Why Kansas City? Because beneath the city lies a thick, nearly level limestone bed with an overlapping cap of impermeable shale. The abandoned limestone mines are 30 to 200 feet below the surface. Support pillars 20 to 30 feet in diameter are left spaced 30 to 40 feet apart. Ceiling heights are 12.5 to 16 feet.

The largest of the warehouses and offices occupying such

space has three miles of well-lighted, underground roads. Mining is still going on, with twenty acres of subterranean space added yearly. Underground space has several advantages. It is inexpensive. Floors will support unlimited weight. There is no noise or vibration. Heating and cooling costs are much less than for surface structures, and it is fireproof.

Kansas City's central location in the United States has made it an ideal place to store food and other items being shipped across the country. Because of the naturally cool temperatures underground, Kansas City has become the world's largest refrigerated storage area. One warehouse can accommodate eighty freight cars at one time at its two underground rail spurs.

One Kansas City company which has gone underground is the Brunson Instrument Company. Brunson manufactured surveying instruments used by astronauts when they landed on the moon. At their former above-ground location, precision settings could be made only between the hours of 2:00 A.M. and 4:00 A.M. when traffic was lightest. In the vibration-free, dust-free underground, settings can be made at any time.

UNDERGROUND STORAGE

One of the world's largest warehouses, Underground Vault & Storage, Inc., is in another kind of mine, a salt mine in Hutchinson, Kansas. Such items as old movies, microfilm records from the city of Los Angeles, X rays from hospitals, and coin collections are stored there. For fifty years, miners removed rock salt from the mine, leaving 30 miles of dry corridors. The salt vein is 40 miles wide, 100 miles long, and 325 feet thick—enough to supply our nation with salt for the next quarter million years.

Underground storage which houses priceless items requires

UNDERGROUND SPACE CENTER

The University of Minnesota has established an Underground Space Center as a research and information center concerning various aspects of underground development and construction. Their offices are located 110 feet beneath the surface in the Civil and Mineral Engineering building. Much of the Center's planning work has involved the development of mined space under Minneapolis and St. Paul, and the University of Minnesota campus.

THE MILITARY UNDERGROUND

The advantages of going underground have not escaped our military planners. Mount Weather, near Washington, D.C., serves as a bombproof headquarters for our president and the military command in case of nuclear attack. This highly classified base is buried deep into the mountain and is complete with streets and sidewalks, offices, houses, and a hospital. Water is furnished by a large, underground lake. Three-inch-diameter steel springs support the eight three-story buildings inside the mountain. More than 900 of these springs, weighing a ton each, cushion the buildings, people, and electronic gear from any possible jarring due to nuclear blasts or earthquakes.

In the West, NORAD (North American Defense Command) has a command center blasted out of a mountain near Colorado Springs. A display screen 1,400 feet below the granite mountaintop is the nerve center that would give the first warning of an attack on North America.

North of NORAD, scattered across the vast farm country of our northern plains states, are 1,000 seldom-noticed, small,

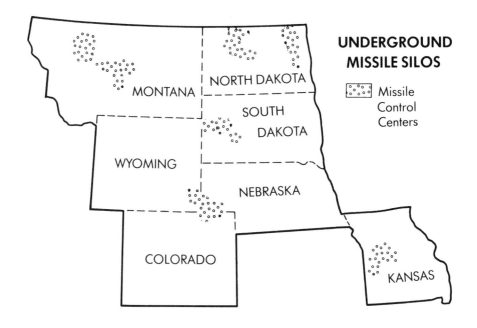

fenced-in enclosures. Most are located on quiet country roads. Each contains a silo, but not to hold grain. These silos are buried in the ground and contain nuclear missiles. One thousand land-based missiles are located in six silo fields spread across seven states. There are 100 underground launch control centers, one for each ten missiles. Most silos contain Minuteman missiles, but some are being reconditioned to take the MX missile.

6
PIPELINES AND PASSAGEWAYS

Pipelines are hollow cylinders through which a fluid material can be transported either by gravity or by the use of pumps placed along its length to keep the fluid moving. Most pipelines are underground, usually placed in a trench several feet deep and then covered over. They remain there unnoticed—unless one happens to see a small sign noting their presence. There are four times as many miles of buried pipelines stretching across America as there are railroad tracks on the surface.

Pipelines carry three types of materials—gases, liquids, and solids. Gas pipelines carry mainly natural gas. Pipelines for liquids carry crude oil or petroleum products—also water for many of our cities. How can solids be carried by pipeline? Solid materials can be transported in the form of *slurries*, which are mix-

Laying a pipeline in trench

tures of liquids and finely ground solid particles. Pulverized iron ore, limestone, and coal are conveyed after being mixed with water. Light materials such as wheat and sawdust are left dry and blown through the lines.

There are 227,000 miles of petroleum pipelines across America. They carry about half of our crude oil and petroleum products. Most of the rest is transported by barge. Five separate crude oil pipelines originate at the Clovally, Louisiana, salt dome storage facility of the Strategic Petroleum Reserve. Stored oil can be taken from eight gigantic storage caverns 1,200 feet below the surface.

MOVING PETROLEUM PRODUCTS

Pipelines transport gasolines, kerosenes, home heating oils, jet fuels, and diesel fuels. Separate products can be transported in batches several miles long, one after the other, in the same pipeline. Computers follow the location of each product on its way to market. There is very little mixing if pressure is maintained. Pumping stations are located from 20 to 100 miles apart. It takes 14 to 22 days for a gallon of gasoline to travel from Houston, Texas, to New York City.

The half-inch-thick steel Trans Alaskan Pipeline carries crude oil across Alaska from north to south. Half of the line is buried underground. There are 1.1 million miles of underground pipelines in the United States for natural gas—about four times the length of those carrying oil.

TUNNELS

Tunnels are mostly horizontal passageways dug through hills and mountains, under cities and waterways. They carry highways, railways, and city subways. They may also be used for water or sewage. Like pipelines, they are sometimes placed or constructed in a trench and then covered over, but tunnels are also bored deep within the earth's crust. Tunnels are excavated through two types of material—rock and earth.

Rock tunnels usually involve blasting. Holes are drilled into the rock face, then packed with sticks of dynamite. After the blast, fumes are sucked out, and the rubble carted away. In most cases a permanent lining of concrete is added. Those tunnels cut through softer but still firm rock, such as limestone or shale, can be dug by tunnel-boring machines. A circular plate covered with disc cutters rotates slowly, cutting into the rock. The roof and walls are held in place by a steel cylinder called a shield, with an overhanging apron attached in the front. Workers remove the earth through doors in the face and install an additional section of permanent tunnel lining of cast iron or preset concrete. When this is completed, the shield is rammed forward and the process is repeated.

NIAGARA POWER PROJECT

Two incredibly large tunnels were constructed in New York State by the Niagara Power Project. The water intake for the power plant is located two and a half miles above Niagara Falls on the upper Niagara River. Water is diverted from the river

Metro subway tunnel in Washington, D.C., under construction. Note the use of a shield.

Construction of water tunnel for New York Power Project

and flows underground to the power plant through two 46-feet-wide and 66-feet-high tunnels. Each of the huge conduits is equal in size to six double-track railroad tunnels. Immense vertical lift gates for each tunnel can block the flow of water, so that either tunnel can be pumped out for inspection or repair.

TUNNELING UNDER WATER

Tunneling through the earth under bodies of water adds the danger of flooding to that of cave-ins. To prevent this, enough air is pumped into the tunnel so that the air pressure exceeds the pressure of the water outside. Workers enter and leave through an air lock with a decompression chamber to prevent "the bends"—a painful condition caused by an imbalance in nitrogen in the body due to the difference in air pressure. Tunnel workers who work under compressed air are called *sandhogs*. The more pressure a sandhog is working under, the fewer hours he can work.

Immersed tunnels are built under water by digging a trench across a river, bay, or other body of water. Steel or concrete tunnel sections, with their ends closed off, are floated into position over the trench and sunk into place. Divers connect the sections and remove the ends. Any water in the tunnel is pumped out. The tunnel is usually covered with earth or sand.

CHESAPEAKE BAY BRIDGE-TUNNEL

A great engineering wonder is the 17.65-mile-long bridge-tunnel across Chesapeake Bay. To keep two existing channels open for large merchant ships and for our Navy's Atlantic fleet based in Norfolk, Virginia, artificial islands were built on either side of the channels. Vehicles leave the bridge, enter a tunnel through one of the islands, travel underground below the channels, and exit to the bridge from the island on the other side. The islands were constructed with sand dredged from the ocean floor nearby, and built into eight-acre mounds rising 30 feet above the surface of the Atlantic Ocean. Huge boulders were placed around the perimeters of the islands to hold the sand in place. Thirty-seven tunnel sections—each as long as a football field—were lowered from barges into trenches and locked to other sections by divers. The completed tunnels were covered with sand.

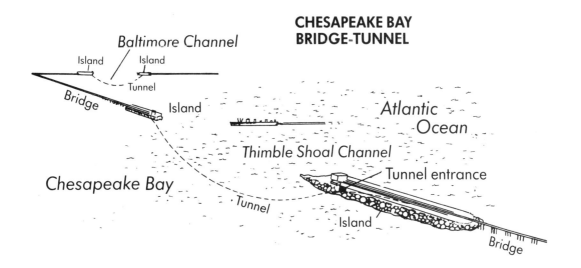

VEHICULAR TUNNELS

Sixty-five miles west of Denver on Interstate 70, one of the world's highest vehicular tunnels (at 11,120 feet) cuts through the Rocky Mountains. Vehicular tunnels have to be much bigger in diameter than railroad tunnels to permit several traffic lanes. The Lincoln Tunnel and the Holland Tunnel are notable vehicular tunnels that connect New York City's Manhattan Island with New Jersey.

Such tunnels present a special problem—how to get rid of exhaust fumes? To overcome this condition, tunnels are designed with massive fresh-air intake ducts below the roadway and massive exhaust ducts along the roof to remove stale air. Fans for the Holland Tunnel under the Hudson River are housed in four ten-story buildings on shore. The air in the tunnel is completely changed every 90 seconds.

RAILROAD TUNNELS

The coming of the railroads brought a great demand for tunnels. By 1850, there were 29 tunnels on American railroads; today there are an estimated 1,000.

One of the first tunnels of importance in America was the railroad tunnel through the Hoosac Range in Massachusetts, built from 1852–1873. In its construction, steam and compressed air drills were used for the first time.

Railroad tunnels through mountains reduce steep grades, allowing locomotives to pull larger loads. The Moffat Tunnel near Denver, Colorado, shortened the route to the West and reduced the altitude a locomotive had to climb by 2,400 feet. It also eliminated many miles of hairpin turns, trestles, and steep inclines.

AQUEDUCTS

An intricate network of underground pipelines and tunnels carries water long distances from lakes and reservoirs to some of our largest cities. The Colorado aqueduct in southern California, which was completed in 1939, carries water through 29 tunnels on its way across the desert from the Colorado River. Other cities with noted aqueducts are San Francisco, Denver, Boston, and New York City. The latter's water supply system is an engineering feat. It consists of three separate systems, all underground, with 18 reservoirs. One tunnel burrows 105 miles through solid rock. Water flows to the city from the higher elevations of upstate New York through gravity.

Watersheds in upstate New York furnish water for New York City residents through a series of aqueducts.

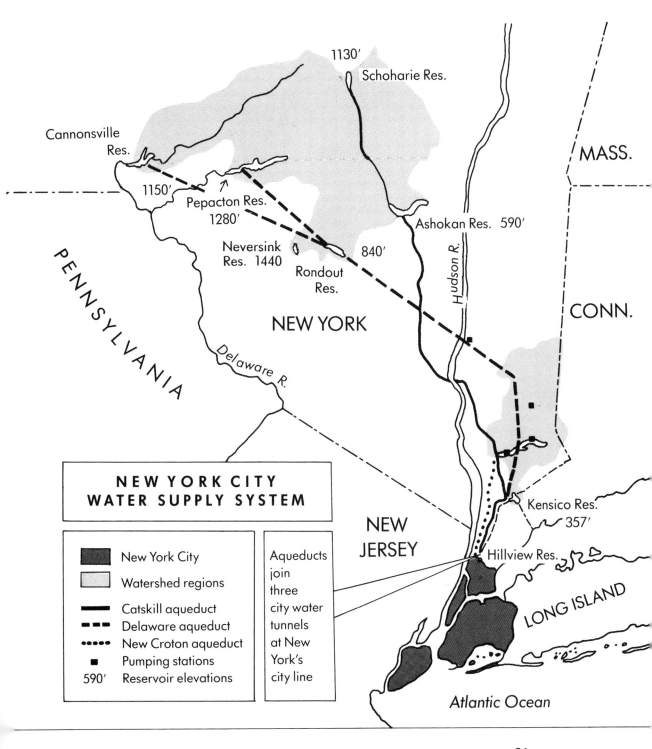

7
UNDER CITY STREETS

Traveling across America by automobile, you may see miles of electric power and telephone lines stretched from pole to pole. But in many places, especially in larger cities, such power lines have been placed underground.

New York City was among the first cities in the country to ban overhead electric and telephone cables. By the 1880s, Manhattan Island was a maze of utility poles and wires. Then, in 1884, it was decreed that all wires on city streets would be taken down and placed in conduits (protective sleeves) underground. No action was taken. But the heavy snow and winds of the Great Blizzard of 1888, which knocked over poles and tangled wires, caused the city to go underground. At about this time other cities also began to place their utilities below the surface.

UNDERGROUND UTILITY LINES

Today, a bewildering labyrinth of utility lines lies beneath our city streets. Workmen dig up the streets to lay ducts, pipes, and conduits through which are threaded the low-tension communications cables—telephone, telegraph, cable television, and connections to fire alarms, police call boxes, and traffic lights. Most utility lines are laid about four feet deep—deep enough so that they will not freeze on cold days, but shallow enough to be accessible in emergencies.

New York City's underground is probably the most congested in America. Of the 3.5 million people who move in and out of Manhattan each working day, more than half travel underground. There are four railroad tunnels, four vehicular tunnels, and thirteen subway tunnels buried under the three rivers surrounding the island. Two major railway terminals with extensive tunnels and underground yards handle rail traffic in all directions.

ELECTRIC AND TELEPHONE LINES

Underground electrical cables and telephone wires are located in protective pipes called *ducts*. High-voltage electric cables and some telephone cables contain insulating oil under pressure. A drop in this pressure indicates a leak. Copper cables for telephone lines are being replaced by newer, lighter, optical fiber cable which provides hair-thin strands of glass that take up much less space, are not affected by water or electrical currents, and can carry almost an unlimited number of telephone calls.

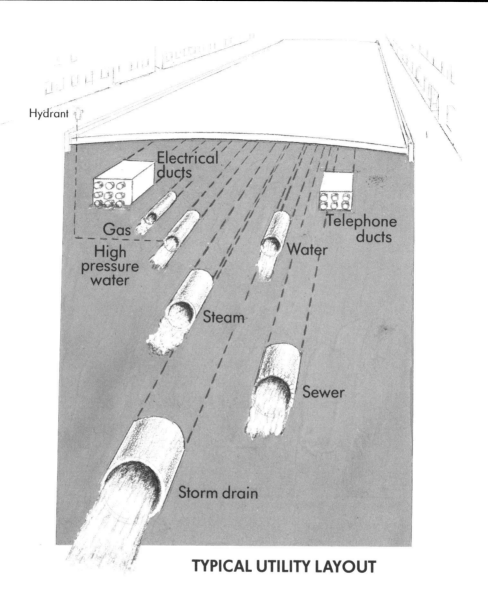

TYPICAL UTILITY LAYOUT

SUBWAYS

A number of major cities have underground subways—Boston, New York, Philadelphia, Baltimore, Washington, D.C., Atlanta, Chicago, and San Francisco. Washington's new subway system carries thousands of tourists and government workers every day.

New York City has by far the most extensive system, with nearly 450 miles of track below ground, the rest on the surface or elevated. In 1985, the New York Transit Authority carried more than a billion passengers on the lines. Subway construction is difficult because of existing sewers, gas mains, steam pipes, and electrical conduits which are beneath the streets. These utilities often have to be moved until subway construction is finished, then replaced before the street is repaved.

NATURAL GAS

Gas flows under pressure through nearly 7,000 miles of gas mains and service pipes. Before the plastic pipes are covered over and buried underground, a copper wire is placed along the top of their entire length, so that workers using metal detectors can locate the pipe if necessary.

STEAM PIPES

Steam is a by-product of generating electricity. As the steam leaves the generators, it is fed directly into street mains. The steam, now at a temperature of 400 degrees Fahrenheit, is used to heat buildings and run machinery.

Steam lines sometimes burst, sending up geysers of hot steam. One of the most damaging of these occurred in 1989. A 24-inch steam pipe exploded with a deafening roar near Gramercy Park in Manhattan. Steam shot up 18 stories high. The blast sent rocks and mud flying in all directions. Three people were killed, twenty-six injured, and thousands were evacuated.

Result of steam pipe explosion near Gramercy Park

SERVICING UTILITY LINES

Everyone has walked over or around manhole covers in the street without giving them much thought, but these are vital doorways to a subterranean world. These small rooms beneath the streets are entered through a circular opening with a cast-iron lid. New York City workers enter the subsurface world through 693,000 manholes. The manholes enable utility companies to repair and maintain their underground equipment. If a manhole has been closed for a long time, fresh air will be pumped into it. Then it is checked for noxious fumes before workers enter.

Street crew pulling cable through manhole

Aging cities of the Northeast have a constant need to replace leaky underground water mains. Philadelphia replaces 20 miles of mains each year. In New York City, water mains, many a century old, burst at the rate of more than one a day, usually at night when the city faucets are shut and pressure builds up. Water supply is disrupted, along with any other utility lines nearby, and subways are often flooded. The city underground requires constant attention. In our larger cities thousands of technicians and laborers are on guard around the clock.

Splicer repairs electric cable inside a manhole.

Bursting water main flooded this subway station.

WATER AND SEWER SYSTEMS

Every city and town is concerned with providing an adequate supply of fresh, pure water, and disposing of sewage and waste in a sanitary and safe way. Underground pipelines and tunnels are employed.

In large cities there are miles of sewer lines carrying waste to treatment plants, other lines for water systems. New York City water flows from high elevations upstate through gravity. When it reaches the city, it flows into water tunnels excavated deep into bedrock under city streets, as much as 800 feet below

New York City's water tunnel No. 3 under construction

the surface. Such tunnels are lined with at least a foot of concrete.

Water tunnel No. 3, recently under construction, began operation in 1990. These city water tunnels are always filled with water. Vertical shafts bring the water up to large trunk mains that distribute it to the five boroughs.

Municipal water systems must be designed to serve much larger demands than the 60 gallons per day per person used by the average family. The New York City system provides a constant supply at high pressure for the 97,000 hydrants used in fighting fires, as well as other needs.

Inspecting completed city water tunnel No. 3

8
IT HAS TO GO SOMEWHERE

Another public service concern in our cities is garbage. Where to put it? It is a nationwide problem. People throw away billions of tons of solid materials each year. It litters our roadsides and pollutes our rivers and lakes. About 1,500 pounds of garbage and trash are produced for each person in the United States each year.

Today about 80 percent of our trash is disposed of by burying it in the ground under layers of soil at sites called "landfills." But within the past decade more than half the landfills in the country have reached their capacity and been closed. New sites are hard to find. Everyone says, "Yes, we need landfills—but not in my backyard."

It is getting so difficult to find places to dispose of garbage underground that it is being shipped from one state to another for burial. More than half of New Jersey's garbage is being transported out of state to landfills in Pennsylvania, Ohio, West Virginia, and even to Michigan.

LANDFILLS

The material dumped in landfills consists of approximately one-third paper and cardboard, one-fifth yard waste (leaves, grass, clippings, etc.), and about one-tenth each of food waste, metal, glass, and plastic. In New York City, 27,000 tons of garbage a day must be disposed of. Most of it goes by barge to Fresh Kills landfill on Staten Island, the largest city landfill in the world. This 3,000-acre site (which was once a valley) now holds a mountain of garbage which could reach 500 feet in elevation soon after the year 2000, making it the highest point along our Eastern seaboard south of Maine.

Refuse in landfills is compacted with bulldozers, and covered with at least six inches of dirt at the end of each day's operation. What is dumped today will be underground tomorrow. Finally, when the area is full, it is buried under two feet of soil—enough to prevent rodents from burrowing into the refuse.

What is to be done? Separating cans, bottles, and paper to be recycled so that they can be used again will help. Leaves, grass clippings, and other yard waste can be composted (broken down into *humus*). But the real answer is more apt to lie in cutting our consumption.

Perhaps you think you do not add much to the problem. What about your ballpoint pen when it runs out of ink? You throw it in the trash where it joins 1.6 million other pens each year. One of our biggest problems is plastic. Some things, such as vegetable matter, paper, bones, and iron, decompose in the ground, but plastic lives on. Unfortunately, we have been replacing cardboard and paper products with plastic, and steel cans that will rust away with longer-lived aluminum cans. Paper takes only two to five months to decompose, but plastic bags take ten to twenty years. Plastic foam cups almost never disappear. Aluminum cans take 80 to 100 years to decompose.

Garbage being bulldozed at a landfill

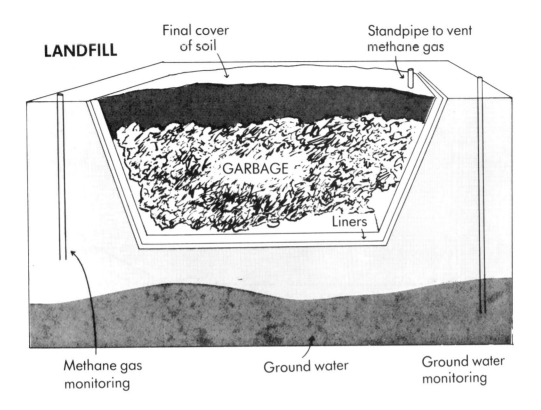

HAZARDOUS WASTE

In 1965, Congress passed the Solid Waste Disposal Act—the first federal law to require safeguards and encourage evironmentally sound methods of disposing of hazardous waste such as acids, toxic chemicals, explosives, and other harmful materials. Landfills for hazardous waste must have at least three impermeable layers of protection. When drums of chemical waste are buried in the ground, they need a plastic, rubber, asphalt, or concrete liner to contain leakage. Below this liner are two additional liners. Observation wells are placed around the perimeter to monitor water quality.

A pressing problem is what to do about leaking storage tanks. More than a half million underground service station tanks are leaking gasoline. Fortunately, only about three percent of the nation's groundwater is now contaminated by such leakage, but groundwater moves slowly and may remain polluted for years or decades.

NUCLEAR WASTE

The Department of Energy has, as a pilot program, constructed an underground storage depot for low-level nuclear waste (such items as contaminated paper, uniforms, or tools) at Carlsbad, New Mexico. Huge storage rooms carved from salt deposits 2,000 feet below the desert will eventually house nuclear waste that will remain highly radioactive for thousands of years. There are still some doubts as to the safety of the site. Under federal law, individual states will be responsible for disposing of their own low-level radioactive waste—an annual total output equal to 400 tractor-trailer loads.

That still leaves high-level nuclear waste, such as spent fuel rods from nuclear power plants and radioactive waste from fifty years of producing nuclear bombs. These materials will remain highly radioactive for 10,000 years. After considerable debate and evaluations of rock environments in many regions of the country, Yucca Mountain, Nevada, was chosen for construction of a repository. Waste will be transported in specially designed casks and disposed of in tunnels 1,000 to 4,000 feet underground. The site will be monitored for any radioactive leaks for many decades.

If all goes according to plan, disposal of high-level nuclear waste will begin at Yucca Mountain in the year 2010. Some critics are concerned that ground water will well up in the mountain over the years, causing explosions and the release of radioactivity. Should that happen, this would not be a safe repository at all.

THE FUTURE

Getting rid of waste is one of the most complex and difficult problems facing the nation in the 1990s. But it is not too late to turn things around. We are learning what to do about the accumulation of so much waste, and have the necessary technology to correct the problems. The completion of this task will have to be carried out by future generations.

PHOTOGRAPHIC CREDITS

Blanchard Springs Caverns, USDA, Forest Service, 33 (bottom right); Carlsbad Caverns National Park, 29, 32 (bottom); Caverns of Sonora, 33 (bottom left); Courtesy of Chevron Corporation, 54; Consolidated Edison of New York, Inc., 67, 68; Department of Environmental Protection, The City of New York (Carl Ambrose), 70, 71; Federal Reserve Bank of New York, 48; Lehman Caves, Great Basin National Park, 32 (top); Luray Caverns, 30, 33 (top right); Glenn McFall, 20; Mammoth Cave National Park Concessions, Inc., 33 (center right); N.Y. Daily News Photo, 66, 69, 74; New York Power Authority, 57; Pacific Gas & Electric Company, 43; Society for Mining, Metallurgy, and Exploration, Inc., 38, 39; Texas Independent Producers & Royalty Owners Association, 37; U.S. Geological Survey (Donald W. Peterson) 24, (H.T. Stearns) 18, (D.A. Swanson) 25, (Photo Library) 21, 23; University of Minnesota, Underground Space Center, 49, 50; Washington Metropolitan Area Transit Authority (Paul Myatt), 56. Drawings and maps are by Christie McFall.

Special thanks to Glenn McFall, who assisted with the maps and drawings; to Sharon Wander and Ralph McFall, who made valuable suggestions; and to my editor, who always keeps me on track.

I am also indebted to James A. Mahaney, P.E., Department of Environmental Protection, City of New York; Paul J. Pasquarello, New York Power Authority; and John Carmody, Underground Space Center.